Light and Sound

Grades 1-3

Written by Isabel Deslauriers, M.Sc., & Tom Riddolls, M.A.C.
Illustrated by Jim Caputo

About the authors:

Isabel Deslauriers is an electrical engineer. She coordinates a science outreach program connecting scientists and engineers with K-12 schools.

Tom Riddolls is a freelance writer, educator, and conservation scientist who has taught science and art on three continents.

ISBN 978-1-55035-881-0
Copyright 2008

Published in the U.S.A by:
On The Mark Press
3909 Witmer Road PMB 175
Niagara Falls, New York
14305
www.onthemarkpress.com

Published in Canada by:
S&S Learning Materials
15 Dairy Avenue
Napanee, Ontario
K7R 1M4
www.sslearning.com

Permission to Reproduce

Permission is granted to the individual teacher who purchases one copy of this book to reproduce the student activity material for use in his/her classroom only. Reproduction of these materials for an entire school or for a school system, or for other colleagues or for commercial sale is <u>strictly prohibited</u>. No part of this publication may be transmitted in any form or by any means, electronic, mechanical, recording or otherwise without the prior written permission of the publisher. "We acknowledge the financial support of the Government of Canada through the Book Publishing Industry Development Program (BPIDP) for this project." Printed in Canada. All Rights Reserved

At A Glance

Learning Expectations	Intro to Light	Light and Society	Spreading and Fading	Shadows	Absorption and Reflection	What is Sound	Humans and Sound	Echoes and Traveling Sound	Pitch and Intensity	Sound Machines
Understanding concepts										
Know how light and sound originate	•	•				•	•			•
Understand how light and sound travel	•					•		•	•	
Understand that humans use light and sound		•					•			
Know that light is energy		•		•						
Understand what shadows and echoes are				•				•		
Understand the characteristics of light (spread, absorption, reflection) and sound (pitch, intensity)			•		•				•	
Know how we hear							•			
Skills of inquiry, design, and communication										
Demonstrate an awareness of the scientific method		•	•		•			•	•	
Classify and sort objects	•	•	•	•	•			•		•
Observe with the senses	•	•	•	•	•	•	•	•	•	•
Work with others	•	•	•	•	•	•		•	•	•
Listen, record, and compare observations	•	•	•	•	•	•	•	•	•	•
Make predictions								•	•	
Record observations, findings, and measurements	•	•	•	•	•	•	•	•	•	•
Communicate the procedure and results of investigations	•	•	•	•	•	•	•	•	•	•
Relating science and technology to the outside world										
Relate scientific concepts to objects in their environment	•	•	•	•	•	•	•	•	•	•
Create new inventions					•					•

Table of Contents

At A Glance™ 2
Teacher Rubric 4
Student Rubric 5
Introduction 6
Materials List 7

Student Activities and Experiments

Introduction to Light
   Teacher Notes 8
   Student Activities 9

Light and Society
   Teacher Notes 16
   Student Activities 17

Spreading and Fading
   Teacher Notes 26
   Student Activities 27

Shadows
   Teacher Notes 34
   Student Activities 35

Transmission, Absorption, and Reflection
   Teacher Notes 39
   Student Activities 40

What Is Sound?
   Teacher Notes 49
   Student Activities 50

Humans and Sound
   Teacher Notes 59
   Student Activities 60

Echoes and Traveling Sound
   Teacher Notes 66
   Student Activities 77

Pitch and Intensity
   Teacher Notes 76
   Student Activities 77

Sound Machines
   Teacher Notes 84
   Student Activities 85

Answer Key 93

Teacher Assessment Rubric

Student's Name: _____

Criteria	Level 1	Level 2	Level 3	Level 4
Understanding Concepts				
Demonstrated understanding of basic concepts of light and sound	Limited understanding	Some understanding	General understanding	Thorough understanding
Demonstrated misconceptions	Significant misconceptions	Minor misconceptions	No significant misconceptions	No misconceptions
Quality of explanations that were given	Shows limited understanding of concepts	Gives only partial explanations	Usually gives complete or almost complete explanations	Always gives complete and accurate explanations
Inquiry, design, and communication skills				
Ability to question, predict, carry out a procedure, observe, and conclude as it relates to the scientific method	Limited ability	Some ability	Good ability	Consistent ability
Use of the correct vocabulary relating to light and sound, and clarity and precision of communication	Limited communication	Some communication	Good communication	Consistent, effective communication
Awareness and use of safety procedures in the classroom	Limited awareness	Some awareness	Good awareness	Consistent awareness
Relating science and technology to each other and the world outside of school				
Demonstrated understanding of the connection between science and technology of light and sound in familiar contexts	Limited understanding	Some understanding	Good understanding	Good understanding in both familiar and unfamiliar contexts
Demonstrated understanding of the connections between science and technology of light and sound and the world	Limited understanding	Some understanding	Good understanding	Good understanding of connections and their implications

Student Self-Assessment Rubric

Name: _____   Date: _____

Put a check mark ✓ in the box that best describes you:

	Needs Improvement	Sometimes	Almost Always	Always
✓ I am a good listener.				
✓ I followed the directions.				
✓ I stayed on task and finished on time.				
✓ I remembered safety.				
✓ My writing is neat.				
✓ My pictures are neat and colored.				
✓ I know what I am good at.				
✓ I know what I need to work on.				
✓ I reported the results of my experiments.				
✓ I discussed the results of my experiments.				
✓ I understand how light works.				
✓ I understand how sound works.				

1. I liked _____

_____

2. I learned _____

_____

3. I want to learn more about _____

_____

Introduction

Teacher Notes

The activities in this book have two intentions: not only to teach students about light and sound, but also to teach general science and technology curriculum objectives through the theme of light and sound.

Throughout the experiments, the scientific method is used. The scientific method is an investigative process which follows five steps to discover if evidence supports a hypothesis:

 Consider a question to investigate.
It is important to choose a question that is clear and that the students are capable of answering. For example, "Does air take up space?"

 Predict what you think will happen.
A hypothesis is an educated guess about the answer to the question being investigated. For example, " I believe air does take up space," A group discussion is ideal at this point.

 Create a plan or procedure to investigate the hypothesis.
The plan will include a list of materials, and a list of steps to follow. It forms the "experiment".

 Record all the observations of the investigation.
Results can be recorded in written, table, or picture form.

 Draw a conclusion. Do the results support the hypothesis? A group discussion is ideal here as well.

The experiments in this book fall under nine topics: **Introduction to Light, Light and Society, spreading and Fading, Shadows, Transmission, Reflection, and Absorption, What Is Sound?, Humans and Sound, Echoes and Traveling Sound, and Pitch and Intensity.** A tenth section on **Sound Machines** provides an excellent way to integrate the sound material that has been learned and introduce a technology component to the curriculum.

In each section, you will find teacher notes designed to provide guidance on the materials and setup needed, as well as highlight any safety precautions.

Materials list

Teacher Notes

Experiments can be done by individual students or by teams.

Each individual/team will need:

Consumables:
- Paper towel tubes
- Batteries for flashlight
- Small amount of salt (2 tbsp)
- Wax paper
- Tissue paper 30 cm x 30 cm
- Tin foil
- Several shoeboxes and larger cardboard boxes
- Popsicle sticks
- Masking tape
- Thin card

Non-consumables:
- Flashlight
- Plastic measuring cups
- Small piece of mirror (about 2 in. or 5 cm square)
- Two identical glasses
- Slinky toy ™
- 4 elastic bands
- Permanent marker
- Cloth 12 in. x 12 in. (30 cm x 30 cm)
- Wood 12 in. x 12 in. (30 cm x 30 cm)
- Foam 12 in. x 12 in. (30 cm x 30 cm)
- Blindfold
- Four baby jars or glasses
- Drumstick
- Scissors
- A small lamp

In addition, you will need (one per class):

Consumables:
- Jug of water or tap
- A pair of bean seedlings or similar vigorous plant

Non-consumables:
- Laser pointer with flat base and on/off switch
- 100W light bulb, clear
- Drum
- Balloon
- Whistle
- Radio, tape deck, or CD player
- Decibel meter
- Thumbtack
- Hammer
- Light socket with cord and switch
- Two thermometers
- Large ball (basketball or soccer ball)
- Sheet of white fabric
- Broom handle
- Tennis ball

Introduction to Light

Teacher Notes

Can You See in the Dark? (page 9)

The students will often insist they can still see when in total darkness. This is due to the continued firing of nerve cells on the retina. Some will even insist they see distinct shapes. The test to prove they are not actually seeing light is to hold an object such as a white ball in front of them, turn out the light, move the ball (e.g., put it on the floor) then ask them if they can see it. They will likely think it is still in front of them. Turn on the light and they will realize they could not see it.

Light Travels Fast (page 12)

Understanding that all light must have a source, this activity enforces the concept that light must move from point A (the source) to point B (i.e., the wall). A difficult concept for young learners is that speeds can be so fast as to seem instantaneous. The purpose here is to teach them that while light does move from A to B, it does so very fast. The speed of light is 186,282 miles per second (300,790 kilometers per second).

Light Travels Straight (page 14)

This principle has been simplified and is technically not 100% accurate. Light travels straight and this is not affected by forces such as gravity, sound waves, and the motion of air. However, light will react to heat energy. Students can be shown this as an advanced step by placing a candle directly below the laser beam (placed close to the source). This will cause the dot of light on the wall to move because the light is bending as it absorbs the additional energy.

Common laser pointers are dangerous to the retina. Students should be cautioned not to point it at faces or to look into the laser pointer.

Introduction to Light

Can You See in the Dark?

Seeing and light go together. Look around you. Where does the light come from? All light has a source. A light bulb is a source. The sun is a source. If you remove the source of light you have darkness.

- a room with no windows and a door
- a towel to block light from under the door
- a tennis ball

With some classmates, close yourselves into a room with no windows. A janitor's closet works well. Identify the source of the light in the room. Then, switch the light off and look around you.

What do you see?

Introduction to Light

Can You See in the Dark?

Name: _____

Worksheet

The brain will "try" to see things in the dark. To test this put the ball somewhere in the dark room with the light on.

predict Will you be able to see the ball with the lights off? _____

observe Switch the light off and have a student move the ball.

What do you see? _____
_____

Where is the ball now? _____

Switch the light on. Where was the ball really?
_____

conclude Can you can see in the dark? _____

If you have no light source, there is _____.

Name: _____

The Source of Light

All light has a source. The sun is a source of light. A light bulb is a source of light.

Look at the pictures below and circle the things that can be sources of light.

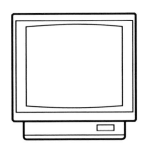  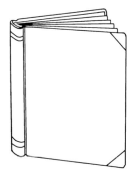

Light Travels Fast!

Light travels from the source to an object.

- a flashlight
- a partner

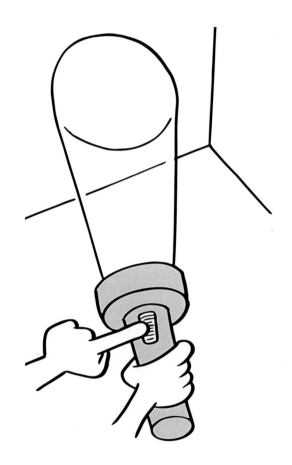

1. Shine a flashlight on the ceiling.

   Light leaves the flashlight and travels up to the ceiling. Just as a car has speed, so does light!

2. Take turns with your partner turning on the flashlight while pointing it at a wall. Watch for the light to leave the flashlight and reach the wall.

Where does the light start?
Where does it finish?
How did it get there?

Introduction to Light

Name: _____

Light Travels Fast!

Worksheet

Light travels very fast! If you were on the Moon with a flashlight and you pointed it at Earth, it would take about one second to get here.

 Will you see the light leave the flashlight and travel to the ceiling? _____

 Turn the light on. What do you see? _____
_____

 Why can't you see the light leave the flashlight and travel to the wall? _____
_____

How fast do you think light travels? _____
Is light faster than a race car? _____

 The speed of light is about 186,000 miles per second (300,000 kilometers per second).

Introduction to Light

Light Travels Straight

Light travels in a straight line. It leaves the light source and travels through the air.

- a laser pointer

Shine a light across the classroom. Without touching the beam of light, try to make the dot on the wall move. Try anything you can to make it move. Noise, wind, another light. Be creative...just don't touch the beam.

Introduction to Light

Name: _____

Light Travels Straight

Worksheet

 Draw what you think will happen to a beam of light from a laser pointer when you try to make it move:

---

observe Draw what happened when you tried to move the light:

---

conclude Can you make the light move? _____

What direction does light travel? _____

 Unless light hits a solid object it will travel straight across the entire universe!

Light and Society

Teacher Notes

You Light up My Life (page 17)
Encourage students to think not about "lights" as objects but light sources. Use examples from the school such as the illuminated "Exit" signs and LEDs on security panels and wrist watches that have lights. This is an exercise that is best accomplished walking around searching for light sources. Be sure to tell them not to list all the ceiling lights in each room and that "ceiling light" only needs one entry.

Light-Detecting Plants (page 18)
Broad-leafed, soft-stemmed plants work best for this. The plants will not need more than a few days, but a fast growing plant such as beans will produce a dramatic effect if left for several weeks as the new branches will all bend straight towards the light source. Both plants should be healthy and vigorous before the experiment; they will fail to respond if they are weak. Expect to see the leaves directed towards the opening in the box. This may be done as a teacher directed activity or in groups of 5-6 students.

How We Use Light (page 20)
There are two principles taught here: that humans use light in most aspects of life, and that light has different functions.

What Is a Light Bulb? (page 22)
Prior to starting this activity, ensure the light switch is in the off position. Encourage the students to think about the conversion of energy. The energy must come from somewhere. Light cannot be produced without an energy source. By explaining the parts of a lamp, including the cord, wall socket, and the wires in the wall, students will begin to understand that the energy does not originate in the light bulb, or in the on/off switch.

Light bulbs get hot. **Don't let students put fingers near socket or inside light housing as there is a risk of electrocution.** When activity is done, put the light bulb out of reach.

Do Light Bulbs Waste Energy? (page 24)
Two identical box set ups may also be prepared with one light not being turned on to introduce the concept of a "fair test" and a "control". For the sheet of plastic, an old overhead works very well. **Never leave the box unattended, the teacher must remove the lamp from the box when the activity is finished (fire and burn hazard).**

Name: _____

You Light Up My Life

Light has many uses. Lights are made in every shape and size from wristwatches to floodlights.

List as many sources of light in your home as you can. Only list each source once! Be sure to check in all the rooms and look for anything that makes light.

(Hint: Things that get really, really hot can also be sources of light.)

_____   _____

_____   _____

_____   _____

_____   _____

_____   _____

_____   _____

_____   _____

How many different sources of light are there in your home? _____

How many are used to see in the dark? _____

How many are used for decoration? _____

Light and Society

Light-Detecting Plants   Teacher Directed Activity

For plants, sunlight is food. Without light, plants would not grow. Plants eat light. If you look at a hungry plant it will show you where it is getting its light.

- two plants
- two cardboard boxes the same size and a little bit bigger than the plants
- black construction paper
- a room with windows in one wall only

1. Cover the inside of the boxes with black paper.
2. Cut off the flaps from the top of one box. Cut off the side panel on the other box.
3. Place one plant in each box and put them as far from the window as possible, with the open side toward the window.
4. Draw the plants in their boxes on your worksheet.
5. Leave the plants for 1 week before looking at them. Draw what you see on your worksheet.

Light and Society

Name: _____

Light-Detecting Plants   Worksheet

predict After a week in the boxes will both plants look the same? _____

observe Draw the plants when they went into the boxes:

Draw the plants after one week in the boxes:

conclude Why did one of the plants need to turn its leaves?

_____

_____

Name: _____

How We Use Light Part 1

Humans make and use a lot of light. In this drawing identify sources of light and color them.

 We produce so much light that big cities like New York, London, and Tokyo can be seen from outer space!

Light and Society

Name: _____

How We Use Light Part 2

Color the sources of light in this picture:

Give three different uses for lights in the house (i.e., to read in the dark) _____

_____

_____

One thousand years ago, everyone went to bed when the sun went down. Now we just turn on a light!

© On The Mark Press • S&S Learning Materials　㉑　OTM-2127 • SSB1-127 Light and Sound

Light and Society

What Is a Light Bulb? Part 1

Name: _____

Teacher-Directed Activity

 you'll need:

- light bulb
- light socket with cord and switch

 what to do

1. Your teacher will screw a light bulb into the socket.

predict Will the light come on? _____

observe Did the light come on? _____

2. Your teacher will plug the light into the wall.

predict Will the light come on? _____

observe Did the light come on? _____

3. Your teacher will turn the switch on.

predict Will the light come on? _____

observe Did the light come on? _____

What Is a Light Bulb? Part 2

Name: _____

Light is energy. You can get light energy from other types of energy, like electricity. A light bulb changes electrical energy into light energy.

1. Can a light bulb light up all by itself? _____

2. Why did the light turn on when the switch was turned on? _____

3. What comes through the cord that the light bulb uses to make light energy? _____

4. In the picture, draw a line to show where the energy in the light bulb comes from. Circle the place where the energy changed from electricity into light.

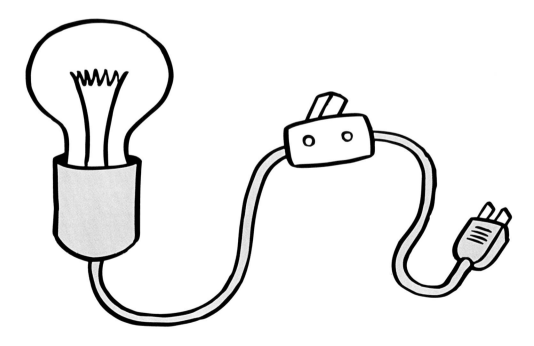

Light and Society

Do Light Bulbs Waste Energy?

Teacher-Directed Activity

Never leave the box unattended. Teacher must remove lamp from the box when activity is finished.

- a small lamp
- a cardboard box **not less** than 1 ft. (30 cm) square, with lid or flaps that close tight
- a thermometer
- a sheet of clear plastic about 9 in. (20 cm) square

1. Cut a small window in the side of the box, about 6 in. (15 cm) square, and tape the sheet of clear plastic over it.
2. Place the lamp in the box together with the thermometer so it is visible through the window.
3. Record the temperature on your worksheet.
4. Close up the box and tape the lid shut.
5. Turn on the light.
6. Every 5 minutes, record the temperature you see on the thermometer. End your experiment after 20 minutes.

Light and Society

Do Light Bulbs Waste Energy?

Name: _____

Worksheet

 What will happen to the temperature inside the box when the light is turned on?

_____

Record your observations:

Time	Temperature
START	
5 minutes	
10 minutes	
15 minutes	
20 minutes	

 How hot did it get inside the box? _____

Besides light energy, what type of energy do light bulbs produce? _____

> We have many light bulbs in our homes. Do you think our houses get warmer by leaving them on?

Spreading and Fading

Teacher Notes

Making a Reflectometer (page 27)
A reflectometer is a simple device used to measure light reflecting off a surface. Our device uses a tube inside of which is a piece of paper with numbers printed on it. The numbers range from gray to very faint. The premise behind measuring is that enough light has to be transmitted through the paper for the number to be visible. In very bright light the faintest numbers will be visible. As the light level decreases, the numbers will get harder to see. If very little light is directed into the tube only the darkest letters will be visible.

The longer the tube the more accurate the results. Toilet paper rolls will also work, but paper towel rolls are ideal. A class project may be to try and make one out of a very long tube, such as a poster or even a carpet tube. These will be very sensitive. But even from very long tubes, the paper gauge should be a consistent 4-8 in. (10-20 cm) from the viewing end.

Does Light Spread Out? Part 1 (page 28)
Students will tend to assume that even though light spreads out, more light is there to fill in the gaps. A more dynamic variation would be to have three students walk away from a source pretending to be a flashlight. Once they get a few feet away, three more students walk along the same path. The principles are that the walking pace is the speed of light which does not change, and that the students are further apart as their distance from the source increases.

Does Light Spread Out? Part 2 (page 29)
The setup and procedure will have to be demonstrated first. Some students will find it difficult to choose one area as high, medium, and lowest in terms of brightness but encourage them to do it as they see it; there is no right answer.

If there is time when students have finished, the exercise could be repeated using a long roll of paper on the floor with markers. In this situation the students will be able to see the light spreading and fading to darkness.

Brightness Decreases with Distance (page 30)
Do this activity in a darkened room. It is important that the number card used in the tube is not printed too dark. Also the faintest number needs to be truly gray and not just speckly black as some photocopiers produce. Use craft paper rather than printed tones on white paper.

How Does Light Spread Out? (page 32)
This analogy of light as sand allows for measuring and graphing.

Spreading and Fading

Making a Reflectometer

A reflectometer measures how bright light is. Let's make one!

- two paper towel rolls
- scissors
- tape
- a partner

1. Cut out the number circle along the dotted lines.
2. Hold the tube steady on the desk.
3. Have your partner press the circle over the top of the tube, fold over the edges and tape them down. Make sure the number side is facing into the tube.
4. Make a cut down the length of the other tube.
5. Slide this cut tube over the first tube so they overlap about 2 inches (8 cm). Tape the cut closed. When looking into the tube, always look into the taped end.

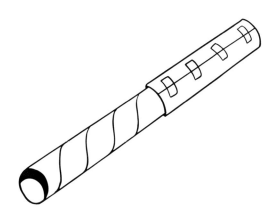

Spreading and Fading

Does Light Spread Out? Part 1

As light travels away from its source, what happens to it? Let's pretend to be light and find out.

- a large space such as a gym or school yard
- your classmates
- a ball

1. Stand crowded together in a tight circle around the ball.

2. Pretend the ball is a light source and you are all light produced by the source.

3. Walk away from the ball in a straight line, just like light traveling from a flashlight.

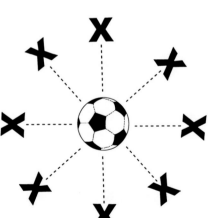

4. After about ten steps try to reach out and touch the "light" closest to you? Can you?

This is what light is like. The closer to a source the more light there is, and the brighter it is. Farther from the source, light is spread out.

Spreading and Fading

Does Light Spread Out? Part 2

Name: _____

- a pencil
- a flashlight
- masking tape
- a large piece of white paper

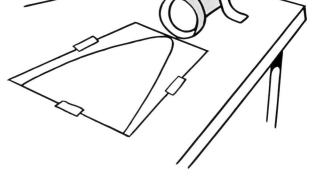

1. Tape the paper to your desk.
2. Lay the flashlight on the desk at one end of the paper and tape it in position.
3. Turn on the light.
4. Trace the outline of the beam of light with your pencil on the paper.
5. Circle what you think is the brightest area and the dimmest area.
6. Put an S next to the source and switch off the light.

 Was the S closest to the brightest area or the dimmest area? _____

What happens to light as it travels away from its source? _____

Spreading and Fading

Brightness Decreases with Distance

Light spreads out when it travels away from the source.

Question: Is light brighter at the source?

 you'll need:

- reflectometer
- a flashlight
- a dark room with heavy curtains
- a flashlight resting on a desk

 what to do

1. Turn on the flashlight.
2. Stand two steps back from the flashlight and look through the tube at the light source.
3. Record the lowest number you can read.
4. Step back two steps and look at the light.
5. Record the lowest number you can read.
6. Continue moving back two steps and record the lowest number until you run out of room.

Spreading and Fading

Brightness Decreases with Distance

Name: _____

Worksheet

?predict? Will there be **more**, **less**, or **the same** brightness of light as you move away from the source?

_____

observe What did you see through the tube as the light was moved farther away?

Number of steps from source	Lowest number I could read
2	
4	
6	
8	
10	

conclude Put a checkmark beside the correct statements in the list below:

_____ I could always read all of the numbers in the tube.
_____ The numbers became harder to read as I moved away from the light source.
_____ Light gets dimmer as it moves away from its source.
_____ Light does not decrease in brightness as it moves away from its source.

Spreading and Fading

How Does Light Spread out?

Light is energy which travels as small bundles. These bundles spread out as they travel.

- your hand
- two measuring cups
- lots of dry sand

1. Outside, bury one measuring cup halfway in the sand. If you do not bury it, the sand may knock the cup over.
2. Pour 1/2 cup of sand into your hands. Stand one step back from the cup and toss in the sand. Record on the worksheet how much sand you caught in the measuring cup. This is how much "light" reached your "object".
3. Pour another 1/2 cup of sand into your hands and take two steps back. Toss the sand into the cup.
4. Continue to toss your 1/2 cup of sand into the measuring cup taking two steps back each time.

Spreading and Fading

How Does Light Spread out?

Name: _____

Worksheet

 **predict** How much of the sand you toss from 12 steps away will land inside the cup? _____

 **observe** Put a checkmark in the column that describes what you see:

Distance from cup	How much sand lands in the cup				
	all of it	most of it	half of it	a little bit	none
1 step					
2 steps					
4 steps					
6 steps					
8 steps					
10 steps					
12 steps					

conclude. What is happening to the sand that makes it hard to get in the cup? _____

What does this tell you about what happens to light as it moves away from the source?

_____

Shadows

Teacher Notes

Changing Shadows (page 35)
This activity explores the nature of shadows. Students will learn that the shape of a shadow is determined by the shape of an object's surfaces turned toward the light. Turning a piece of paper will cause a small shadow if held with its thin edge to the light and a large shadow if held with its flat surface toward the light. Rotating a ball will show no change in its shadow.

The objects may be gathered by the teacher and collected in a box for the students to pick from, assigned to the groups by the teacher, or selected by the students from objects in their desks or backpacks. Objects cannot be bigger than the piece of paper.

It is easier for the students to see the shadows if the light levels in the room are lowered.

Students should work in three's around one desk. One to hold the flashlight, another to hold the object, another to draw and label the shadow. The person who is drawing chooses the position the object is held.

Shadows and Light (page 37)
The goal of this exercise is to have the students understand that the shape of a fixed object's shadow (in this case a block representing a building) is determined by the position of the light source. In the morning and in the evening, the sun casts longer shadows than at mid-day. The activity can be made more elaborate by using a classroom-made model of a city, town, or farm.

Changing Shadows

When light is blocked by an object, you get a shadow.

Is an object's shadow always the same shape?

- a flashlight
- six objects of different shapes from your teacher
- six pieces of blank paper and a pencil
- two teammates

1. Pick up an object.
2. One student holds the flashlight above the desk, as high as he or she can while standing on the floor.
3. Another student will take an object and hold it above the paper.
4. Another student will draw the outline of the object's biggest shadow and the object's smallest shadow onto the piece of paper.
5. Label your paper.
6. Switch places with your teammates and repeat with the other objects.

Name: _____

Changing Shadows

Worksheet

Which object will produce the biggest shadow? _____

Which object will produce the smallest shadow? _____

Which object did produce the biggest shadow? _____

Which object did produce the smallest shadow? _____

Why did the shape of the shadow of some objects change? _____ _____

Why did the shape of the shadow of some objects not change? _____ _____

What makes a shadow? _____ _____

Shadows and Light

A shadow is the shape created when an object blocks light. The sun makes shadows when it shines on objects like trees, buildings, and people.

What happens to shadows when the light changes direction?

- three to six blocks
- a flashlight
- paper and pencils

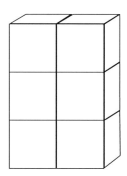

1. Your blocks represent a building.
   Place them on the paper to make a building.
2. Predict on your worksheet what will happen when you shine your flashlight onto the building from above.
   This is the sun at high noon.
3. Predict on your worksheet what will happen when you shine your flashlight onto the building from the side.
   This is the sun at sunrise and sunset.
4. Shine the light straight down at your building.
   Draw the shape of the shadow on your worksheet.
5. Shine the light from the side onto your building.
   Draw the shadow on your worksheet.

Shadows

Name: _____

Shadows and Light

Worksheet

Question: What happens to shadows when the source of light changes direction?

Guess the shape of the shadow of your building:

Light shines from the top	Light shines from the side

Draw the shape of the shadow of your building:

Light shines from the top	Light shines from the side

conclude
1. Can the shape of a shadow be changed by moving the light source? _____
2. During the middle of the day when the sun is high in the sky, a building's shadow will be _____.
3. First thing in the morning when the sun is low in the sky, the shadow will be _____ _____.

Transmission, Absorption, and Reflection

Teacher Notes

I See Through You! (page 40)
The ability to read through the material is the best test for transparency/translucency. Egg shells, orange juice, and some plastics are everyday items that are translucent. Some of these, such as the egg shell, may come as a surprise. The difficulty will be the inability for students to put objects into a definite group considering that many objects change state depending on things like temperature and thickness. Egg whites, for example, are transparent when fresh and translucent when cooked.

It's Play Time! (page 41)
This activity provides ample opportunity to explore the properties of transparent, translucent, and opaque materials. The characters are opaque and cast shadows on a translucent screen. It may be necessary to bring in a clear shower curtain to show how a transparent material would not work while a translucent one does. The worksheet can be used to draw out understandings of these terms and how they relate to real world applications. If desired, the play can be used as assessment by asking students to dramatize all they've learned about light.

To be effective the room needs to be quite dark. The light should be very bright, such as a bare, unfrosted 100W bulb. The screen could be hung across a broom handle or yard stick suspended across two desks with the audience sitting on the floor.

Bouncing Light (page 43)
The two objectives are to have the students anticipate and witness that light reflects at an expected angle, for example light striking at 45 degrees will be directed away at 45 degrees. Light directed straight at the mirror will be reflected directly out from the mirror. The activity requires steady hands and some fine adjustment. It may be of benefit to the student to watch the teacher demonstrate the basic set up first.

What Reflects Light? (page 45)
Gray paper is not critical, colored paper can be used. Also a flashlight is not necessary as the pieces of paper can be set up in front of a window. An extension to this experiment is to test the transmission of the paper using the reflectometer tube.

Absorbing Light (page 47)
This can be a class activity with one set of boxes or can be done by groups of students. Almost any size box will work, just make sure they are the same size; two shoe boxes would be perfect. Real glass will provide better results and this is possible if only one set of boxes are made. This activity can be extended to measure the sun's heat over a long period, or with a variety of weather conditions, and indoors and out.

I See Through You!

Name: _____

Light passes through some things and we can see right through them, like glass. We call these things **transparent**. A test to see if something is transparent is to see if you can read through it. If you can see the words, the material is transparent.

Some things let some light pass through, but keep or absorb some light, like a piece of paper. Hold this page up to the window. Some light will pass through it. We call things that some light passes through **translucent**.

Some things don't let any light pass through, like your hand. We call things that block light **opaque**. Hold your hand up to the window. No light will pass through it.

List some transparent, translucent, and opaque objects found in your classroom:

Transparent	Translucent	Opaque

Transmission, Absorption, and Reflection

It's Play Time!

- a strong light source
- a thin white cotton sheet
- Popsicle sticks
- stiff cardboard, paper bags, yarn
- tape and scissors

Write a short play about light. Create characters for your play by cutting out cardboard and paper shapes and taping them to sticks. The actors sit between the light and the screen and the audience sits on the other side of the screen.

Transmission, Absorption, and Reflection

Name: _____

It's Play Time!

Worksheet

1. Circle the word that best describes the sheet:

   shadow   transparent   light source   translucent   opaque

2. Circle the word that best describes the characters on sticks:

   shadow   transparent   light source   translucent   opaque

3. Circle the word that best describes what the audience watches on the screen:

   shadow   transparent   light source   translucent   opaque

4. Do the figures on the screen look different than the cardboard cutouts? How? _____
   _____

5. What type of material is best used for a shadow puppet screen? _____

Transmission, Absorption, and Reflection

Bouncing Light

Light bounces off smooth surfaces just like a ball does. If you throw a ball straight down it will come straight back up. If you throw a ball at the ground away from you it will not bounce back towards you. Try it!

you'll need:

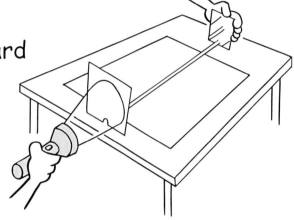

- a large piece of white paper
- a small piece of stiff cardboard
- a pair of scissors
- a small mirror
- a flashlight
- a partner

what to do

1. Cut a small hole out of the side of the cardboard.
2. Lay the sheet of paper on your desk and hold a mirror still at one end.
3. Shine the light through the hole in the cardboard and point it at the mirror.
4. Line the mirror and cardboard up so that the light shines directly back into the hole in the card.
5. On the paper draw a line where the light shines.
6. Without moving the mirror, move the light so it hits the mirror at an angle.
7. Draw the line of light with your pencil.

Transmission, Absorption, and Reflection

Name: _____

Bouncing Light

Worksheet

 Can light bounce off objects? _____

 In the space below, draw the light beam:

Light pointing straight 	Light pointing at an angle 
Mirror	Mirror

 Does light bounce? _____

Bouncing light is called reflection.

Transmission, Absorption, and Reflection

What Reflects Light?

Light bounces off mirrors. Mirrors reflect light. Do all objects reflect light?

- several pieces of paper ranging from white to gray to black, labeled 5 to 0 (white being 5 and black being 0)
- reflectometer tube
- a flashlight
- a partner

1. On your desk lay out the pieces of paper that range from white to black.
2. Hold the tube about 2 in. (5 cm) above the paper. Point the tube at the paper and look through it.
3. Your partner will shine the flashlight on the paper.
4. Record the lowest number visible on your worksheet.
5. Do the same with all the pieces of paper, then switch places with your partner.

Transmission, Absorption, and Reflection

Name: _____

What Reflects Light?

Worksheet

 Which color of paper reflects the most light?
_____

observe Write down the lowest number you could read in the reflectometer:

Paper	What number did you see?
Black 0	
Gray 1	
Gray 2	
Gray 3	
Gray 4	
White 5	

conclude Which piece of paper reflected the most light?
_____

Why can you see more numbers through the reflectometer when you point it at the white paper? _____

_____

© On The Mark Press • S&S Learning Materials — OTM-2127 • SSB1-127 Light and Sound

Transmission, Absorption, and Reflection

Absorbing Light

Light energy gives off heat energy. Measure the heat energy given off by the sun by trapping it in a box.

- two cardboard boxes the same size
- two thermometers
- two pieces of plastic wrap
- black and white paint or paper

One box will be your white box, this reflects light.
The other box is your black box, this traps light.

1. Color the inside of one box white and one box black.
2. Place a thermometer in each.
3. Cover the top of the box with the clear film.
4. Record the temperature before you place the box in the sun.
5. Place the box in a sunny spot and record the temperature after 1 hour.

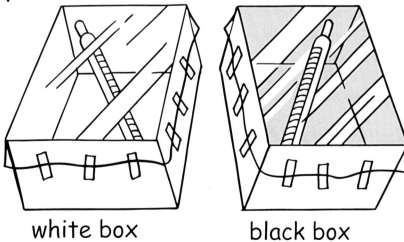

white box   black box

Transmission, Absorption, and Reflection

Name: _____

Absorbing Light

Worksheet

 What will happen to the temperature in the black box when placed in the sun? _____

_____

 What is the temperature inside the boxes before placing them in the sun? _____

What is the temperature inside each box after 1 hour? White box: _____ Black box: _____

Which color box had the highest temperature? _____

 What type of energy was measured inside the black box? _____

How do you know? _____

 What would happen if you put the boxes out on a cloudy day?

What Is Sound?

Teacher Notes

Sound is usually produced by compressing or "squishing" air. This can happen in three basic ways:

1. **Vibration**: When an object vibrates, it intermittently compresses the air around it and produces a sound. For example, a drumstick hitting a drum makes the drum vibrate, or strumming a guitar makes its strings vibrate.

2. **Explosions:** Explosions rapidly compress the air around them and produce a sound like a "bang" or "pop". A balloon that pops and fireworks are examples of explosions that produce sound.

3. **Flow of air:** When air flows against surfaces (for example, when we blow in a whistle or the wind blows against window sills) a whistling sound is produced.

Where Does Sound Come From? (page 50)
Students learn about the different ways sound is produced. The experiment can be done as a group demonstration, with the students completing the worksheets individually, and reinforcing the concept of vibration in **Make A Kazoo** (page 53).

Does Sound Travel? (page 55)
This activity demonstrates how sound travels through the air once it is produced. The sound vibrations can be transmitted from one place to another, or from one object to another.

What Is Sound?

Where Does Sound Come From?

Name: _____

Worksheet

- drum and drumstick
- salt
- balloon
- whistle

Vibrations produce sound.

1. Sprinkle the salt on the drum.
2. Hit the drum with the drumstick.

 What happens to the salt?
_____
_____

What Is Sound?

Where Does Sound Come From?

Name: _____

Worksheet

Blowing air produces sound.

1. Blow in the whistle.

What happens?

_____

_____

Explosions produce sound.

1. Blow up the balloon.
2. Pop the balloon with your foot.

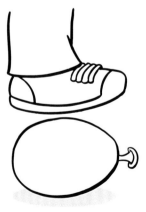

What do you hear?

_____

_____

© On The Mark Press • S&S Learning Materials OTM-2127 • SSB1-127 Light and Sound

What Is Sound?

Where Does Sound Come from?

Name: _____

Worksheet

 What did you discover? Three ways to produce sound! Draw a picture of each one:

Vibration creates sound

Explosions produce sound

Blowing air produces sound

What Is Sound?

Name: _____

Make A Kazoo

 you'll need:

- plastic comb
- about 3 in. x 2 in. (10 cm x 5 cm) of each:
    - ☐ wax paper
    - ☐ newspaper
    - ☐ plastic bag
    - ☐ tissue paper

 what to do

1. Pick a material and fold it over the middle of the comb.
2. Place your mouth over the paper on the toothed side.
3. Hum into the paper (like you would hum into a kazoo).
4. Try the other materials.
5. Answer the questions on the worksheet.

Name: _____

Make a Kazoo

Worksheet

 What will happen when you hum into the comb?

_____

 What happened when you hummed into the comb?

_____

Which material made the loudest noise?

_____

Which material made the softest noise?

_____

 Circle the right answer:

The kazoo makes sounds because the material :

    screams   vibrates   rolls up

How did the different materials change the sound of the kazoo?

_____

_____

_____

What Is Sound?

Name: _____

Does Sound Travel?

 **you'll need:**

- two identical glasses
- jug of water

 **what to do**

1. Fill one glass halfway with water.
2. Dip your finger in the water.
3. Rub the rim of the glass with your finger. Go around and around.

 1. What do you hear? _____

_____

2. What do you see? _____

_____

What Is Sound?

Name: _____

Does Sound Travel?

Part 2:

4. Place the second glass about 4 in. (10 cm) away from the first glass. Fill the second glass halfway (exactly like the first glass).
5. Dip your finger in the water.
6. Rub the rim of the first glass with your finger. Go around and around.

 Does the second glass make a sound? _____

What do you see in the second glass? _____
_____
_____

True or false? Vibrations and sound go together _____

What Is Sound?

Name: _____

Does Sound Travel?

Worksheet

Draw a picture of the two glasses of water.
Draw your finger rubbing the first glass.

[ ]

Your finger makes the first glass vibrate. It produces a sound. This sound travels through the air. It makes the second glass vibrate. Because of this, the second glass also makes a sound.

In your picture above, draw an arrow in the direction the sound is traveling.

How Does Sound Travel?

Name: _____

 you'll need:

- Slinky™ toy
- a friend

 what to do

1. Sit on the ground.
2. Ask your friend to sit in front of you.
3. Stretch the Slinky™ between you, on the ground. Each person holds one end.
4. Swing the Slinky™ from side to side while your friend holds his end still.

 What happens to the Slinky™ at your friend's end?
_____
_____

conclude How is this like sound? _____
_____

The vibration travels from you to your friend. This is how sound travels through the air.

Humans and Sound

Teacher Notes

Humans use sound to communicate, and in this section, the students explore how we produce and hear sounds. Through these activities, students will learn about vocal cords and the eardrum, which, respectively, allow us to talk and hear.

How Do Humans Make Sound (page 60)
Students will sing and use their fingertips placed on their throats to discover that it is vibration that produce singing and talking sounds.

Make an Ear (page 61)
Students will make a model of an eardrum and see how it vibrates when sound reaches it.

How Do Humans Hear Sound? (page 63)
Students will be made aware of basic ear anatomy, and relate this to what they have learned about how sound travels and how it is related to vibration. Thus this should be done after **Does Sound Travel** and **How Does Sound Travel.**

Why Do We Have Two Ears? (page 64)
Students will learn that having two ears makes it easier to determine where sounds are coming from.

These experiments could be an opportunity to discuss hearing safety. Loud sounds damage the eardrum and may even rupture it. Some basic hearing safety rules that can be emphasized include:

- Don't listen to the radio or TV too loudly.
- When listening to music with earphones, if you cannot carry on a conversation with someone next to you, the music is too loud!
- Always wear ear protection when working around noisy tools.

You can tell your students that they will have the opportunity to form a "sound patrol" (page 82) to try to identify sources of dangerous sound in your school!

Humans and Sound

How Do Humans Make Sound?

Name: _____

You know that vibrations produce sound. How do humans make sound?

1. Touch your fingers lightly to your throat.
2. Say "ahhh" like at the dentist, or sing a song.

 Using your senses, make observations. What happens? _____

_____

Humans make sounds using vibrations. The vocal cords in your throat vibrate when you say "ahhh". Color the picture and make the vocal cords pink.

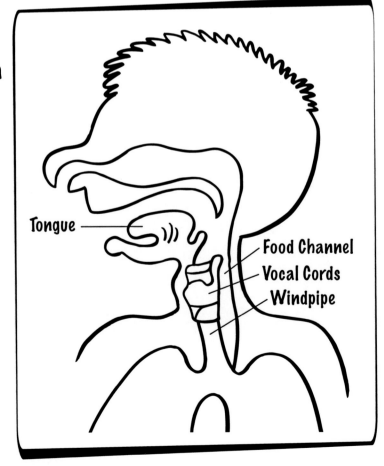

© On The Mark Press • S&S Learning Materials   OTM-2127 • SSB1-127 Light and Sound

Humans and Sound

Make an Ear

- a glass
- a piece of wax paper that can cover the glass opening
- elastic
- permanent marker
- salt

1. Write "eardrum" on the wax paper.
2. Cover the glass opening with the wax paper.
3. Fasten the wax paper over the glass with the elastic.
4. Sprinkle the salt on top of the wax paper.
5. Say "ahhh" over the glass. Try other sounds.

Humans and Sound

Name: _____

Make An Ear

Worksheet

 What happened to the salt on the wax paper when you made a sound over the glass?

_____

_____

The same thing happens in your ear.
The eardrum inside your ear vibrates when you hear sounds.

Color the picture and make the eardrum blue.

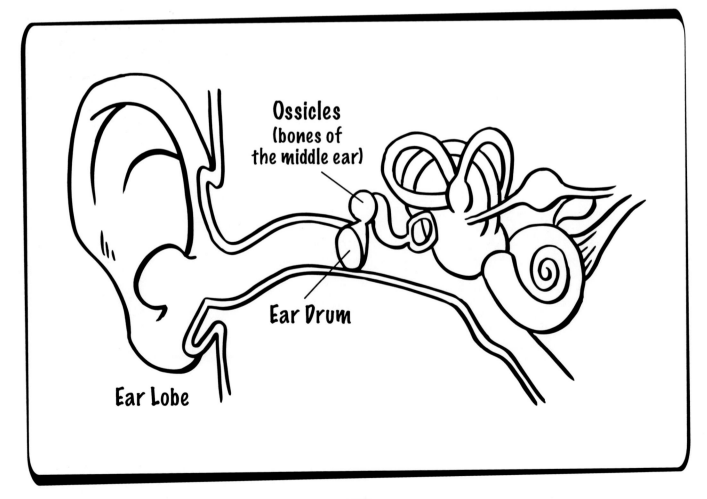

Name: _____

How Do Humans Hear Sounds?

Inside each ear, we have an eardrum. When a sound hits the eardrum, it vibrates. This sends a message to our brain, and we know we heard a sound.

Sound travels from the horn to the eardrum. In the picture below, color the eardrum blue. Draw a red arrow from the horn to the eardrum. Color the brain pink. Draw a green arrow from the eardrum to the brain.

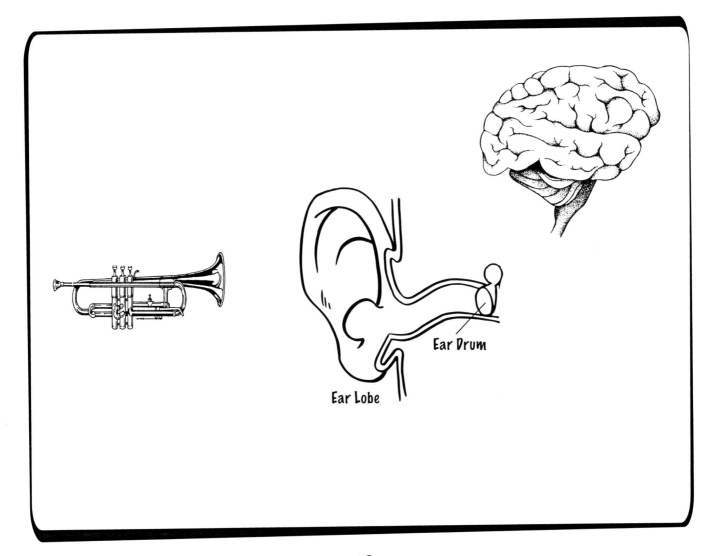

Humans and Sound

Why Do We Have Two Ears?

Teacher-Directed Activity

Why do you think we have two ears?

1. The class sits in a circle.
2. One student sits in the middle. He closes his eyes.
3. The teacher points to a student. She makes a sound.
4. The student in the middle tries to guess who made the sound.
5. After five tries, repeat the activity while the student in the middle covers one ear.

Humans and Sound

Why Do We Have Two Ears?

Name: _____

Worksheet

 Draw a picture of what you did.

[ ]

conclude — Was the activity easier with one or two ears? Circle your answer.

| **With two ears** | **With one ear** | **The same** |

Compare with your classmates.
What did they say was easier?

| **With two ears** | **With one ear** | **The same** |

Why do you think we have two ears?

_____

_____

Echoes and Traveling Sound

Echoes and Traveling Sound

Teacher Notes

Speed of Sound (page 67)
In this activity, the class goes to a large open space such as a football field. The teacher chooses a student to walk to the other end of the field, blow a whistle and drop his hand at the same time. The remaining students will observe that sound doesn't travel instantaneously as they will see the hand drop before hearing the whistle. A worksheet is provided for students to complete individually. A class discussion on how fireworks are heard a few moments after being seen could follow this activity. Also, you can explain to the students that they can calculate the distance to an electrical storm by counting the delay between seeing and hearing the lightning (distance in miles = number of seconds divided by 5, distance in kilometers = number of seconds divided by 3).

Sound Travels (page 69)
In this activity, the teacher places a radio in the middle of the class, speaker pointing upwards, and students sit around it. The students will observe that everyone in the circle can hear the radio, thus showing that sound travels in all directions.

Blocking Sound (page 71)
In this activity, the students will learn that different materials can block sound more or less effectively. The experiment follows the scientific method (see introduction) by asking the students to predict which materials will block sound best, then carrying out an experiment to confirm or disprove their prediction. During the experiment, the teacher places a radio in the middle of the class, speaker pointing upwards, and students sit around it. The teacher chooses students, one at a time, to place one of the materials on top of the speaker. The group then records their observations individually: did the material block the sound effectively? A follow-up class discussion could include questions such as: What would be a good material to construct a soundproof room? What would be a good material to make earplugs?

Echoes and Bats (page 73), Bat and Mosquito Game (page 74), Would You Be a Good Bat? (page 75) Echoes and echolocation are addressed in these activities. Some animals, such as bats, survive in dark areas and thus must rely on means other than sight to locate their prey. Echolocation consists of making small noises and listening to the echoes produced to find prey. A text illustration (page 73), a role-playing activity (page 74), and an experiment (page 75) are suggested, in order to appeal to students who learn in different ways.

Echoes and Traveling Sound

Speed of Sound

- a whistle
- a large outdoor space (e.g., football field)
- a friend

1. Go outside and stand at one end of the outdoor space.
2. One student walks to the opposite end of the space with the whistle (as far as he can go).
3. The student raises his hand.
4. The student blows the whistle and drops his hand at the same time.
5. What happens first? Do you hear the whistle first or do you see the hand drop first?

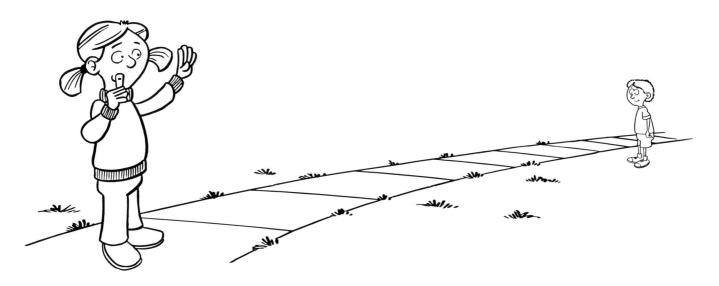

Echoes and Traveling Sound

Name: _____

Speed of Sound

Worksheet

Cut out the sentences at the bottom of the page. Glue them in order.

1. _____

2. _____

3. _____

4. _____

She blew the whistle and dropped her hand.

One student walked to the other end of the field.

We saw the hand drop.

We heard the whistle.

Echoes and Traveling Sound

Sound Travels

Teacher-Directed Activity

- a radio

1. Place the radio in the middle of the room. The speaker should point up.
2. Sit in a circle around the radio.
3. Turn on the radio.
4. Who can hear the radio? Raise your hand.

Sound Travels

Name: _____

Worksheet

Circle your answers.

1. What do you think is true?

   a) Sound travels in all directions

   b) Sound travels in only one direction

2. Who could hear the radio?

   a) Everyone in the classroom

   b) Only some people

3. Draw a picture of the experiment:

4. Add arrows showing how the sounds traveled.

Echoes and Traveling Sound

Blocking Sound

Teacher-Directed Activity

- a radio
- materials big enough to cover the speaker: cardboard, tissue paper, cloth, wood, foam

1. Place the radio in the middle of the room. The speaker should point up.
2. Sit in a circle around the radio.
3. Turn on the radio.
4. Guess which material will block sound the best. Record your guess.
5. A student puts one of the materials on the speaker.
6. Does the material block the sound?
7. Try the other materials on the speaker.
8. Do they block the sound?

Blocking Sound

Name: _____

Worksheet

 **predict** Which material will block the sound best?
This is what I guess: _____

observe Put a checkmark ✓ in the column that tells what happened:

Material	Did not block	Blocked a bit	Blocked very well
Wood			
Foam			
Cardboard			
Tissue Paper			
Cloth			

conclude The material that blocked the sound best was:
_____

Echoes and Bats

Name: _____

Did you know that sound bounces off walls and objects? We call this an "echo".

Bats use echoes to find insects. They make a small sound. If the sound bounces back, they know they have found an insect.

Draw a picture of a bat using echoes to find insects:

Echoes and Traveling Sound

Bat and Mosquito Game

- a blindfold

How to play the game:

1. One student is the bat. He is blindfolded.
2. Two other students are the mosquitoes.
3. The rest of the class holds hands in a circle. The bat and mosquitoes are in the middle.
4. The bat says "beep-beep".
5. The mosquitoes send a pretend echo by saying "bop-bop".
6. The bat tries to catch the mosquitoes. He walks in the direction of the echo. The mosquitoes try to escape.

Echoes and Traveling Sound

Would You Be a Good Bat?

Name: _____

- a book
- a blindfold
- a friend

1. Put on the blindfold.
2. Your friend puts the book 6 in. (15 cm) in front of your face.
3. Talk and listen to the echo.
4. Your friend removes the book.
5. Talk. There is no echo.
6. Now your friend will decide when to put the book in front of you. Talk, and try to guess when the book is in front of you by listening to the echo.

observe Was it easy to guess if the book was in front of you? _____

conclude How is this like a bat catching mosquitoes? _____

Pitch and Intensity

Teacher Notes

Sounds can be characterized by two different attributes: pitch and intensity.

PITCH

High-pitched sounds are "high notes" and include squeaky noises and sounds made by a bell, or a triangle. Low-pitched sounds are "low notes" and include sounds made by a bass drum, or a rumble of thunder.

You could use a musical instrument to introduce the concept of high-pitched or low-pitched sounds (high notes and low notes, respectively) to your students. Once they understand the concept, you can reinforce it by playing two notes and asking which one is the highest or the lowest.

Make a Xylophone (page 77)
Students will build their own musical instrument (a xylophone with baby jars filled with various amounts of water) and the concept of pitch can be reinforced. The baby jars containing less water will give higher notes, and the baby jars with more water will give lower notes.

INTENSITY

Intensity refers to the loudness of sounds.

The Intensity of Sound (page 80)
Students will learn to use a decibel meter to measure how loud a sound is. (If you cannot obtain a decibel meter, you could conduct these experiments by relying on the students' perception.) The students will discover that the louder they turn the volume of the radio, the larger the reading on the decibel meter – therefore, the decibel meter measures how loud a sound is. They will then use the decibel meter again and find out that sound becomes softer as they move away from a sound's source. The subject of standing too close to loudspeakers or noisy tools could be brought up at this point.

Sound Patrol (page 82)
A sound patrol will be formed and the students will search for potentially dangerous sources of loud noise around their school. The experiment follows the scientific method (see introduction) by asking the students to predict which sources will be the loudest, then carrying out the measurements to confirm or disprove their prediction.

Pitch and Intensity

Make a Xylophone

Sounds have different pitches. A screech and a meow are high-pitched sounds. The 'ahhh' you make at the doctor and a cow mooing are low-pitched sounds.

- four or more glass baby food jars
- water
- a drumstick or other stick

1. Fill each jar with a different amount of water.
2. Bang the drumstick on each jar.
3. Listen to the sounds.
4. Put the jars in order from lowest to highest pitch.
5. Can you invent a song with your xylophone?

Pitch and Intensity

Name: _____

Make a Xylophone

Worksheet

 Draw a picture of your ordered baby jars:

```
┌─────────────────────────────────────────────┐
│                                             │
│                                             │
│                                             │
│                                             │
│                                             │
│                                             │
└─────────────────────────────────────────────┘
```

conclude — Circle the correct answer.

The lowest pitched sound came from the jar with:

| the most water | the least water |

The highest pitched sound came from the jar with:

| the most water | the least water |

Pitch and Intensity

Name: _____

How Loud Is Sound?

Sounds can be loud or soft.
Intensity tells us how loud a sound is.

- a decibel meter
- a radio

1. Turn on the radio at low volume.
2. Take two big steps away from the radio. Hold the decibel meter. What is the number on the meter? _____
3. Put the radio on medium volume. Take two big steps away from the radio. Hold the decibel meter.
   What is the number on the meter? _____
4. Put the radio on high volume. Take two big steps away from the radio. Hold the decibel meter. What is the number on the meter? _____

conclude. What does a decibel meter measure?

_____

Pitch and Intensity

Name: _____

The Intensity of Sound

 you'll need:

- a decibel meter
- a radio

 what to do

1. Turn on the radio at medium volume.
2. Take two big steps away from the radio.
   Hold the decibel meter.
   What is the number on the meter? _____
3. Take two MORE big steps away from the radio.
   Hold the decibel meter.
   What is the number on the meter? _____

 What happens when you move the decibel meter away from the radio? _____

_____

 Why does this happen? _____

_____

Pitch and Intensity

Name: _____

At the Zoo

Each pen is for an animal that makes a certain kind of sound. Draw a picture of an animal that makes that sound.

Pitch and Intensity

Sound Patrol

- a decibel meter
- a sound patrol disguise (optional) : dark glasses, badge

1. Cut out the sound patrol badge below.
2. Choose 5 sound sources around your school. Fill in the first column of the table.
3. Which do you think will be the loudest? Record your guess.
4. Test 5 sound sources around your school with the decibel meter.
5. Record your findings on the worksheet.

Pitch and Intensity

Name: _____

Sound Patrol

Worksheet

 **predict** Which sound source will be the loudest?
This is what I guess: _____

observe This is what happened:

Sound source	Decibel meter reading

conclude The loudest sound source was: _____

 **to think about...** How can you reduce sound levels in your school?

Sound Machines

Sound Machines

Teacher Notes

In this section, students will use what they learned about sound (how vibration creates sound, how sound travels from one place to another, how air flow creates sounds, and how sound is characterized by pitch) to build and understand machines and instruments used in everyday life.

How Does a Speaker Work? (page 85)
Students will see how a speaker inside a radio vibrates to create sound. Before this activity, you may want to remind your students of the three ways sounds can be created (vibrations, explosions, and air flow) and ask them to guess which one is used by a radio speaker.

Make sure the students do not touch any exposed wires inside the radio. Cover the wires with electrical tape if necessary.

Make a Tin Can Telephone (page 86)
Students will build a tin-can telephone. You can first review the **How Does Sound Travel?** activity, which explained how sound travels. With a tin can telephone, a student talks into a tin can, making the can vibrate; the vibrations from the can are transferred to the string, and in turn to the other can. When the other can vibrates, the second student hears the message like in a real telephone. You can tell your students that this is how real telephones work, except that in real telephones, electricity is used to transmit the vibrations from one end to the other.

To prepare the telephones, cover the edges of the cans with duct, electrical, or masking tape. Hammer the thumbtack into the bottom of the cans to make one pinhole in each. Tin cans can be reused.

Make a Pan Flute (page 87) and Make a Guitar (page 89)
Students will make musical instruments. These two activities are good synthesis activities as they revisit the concepts of vibration creating sound and one of the characteristics of sound, pitch. For **Make a Guitar**, if you are using thin elastics, have the students tie knots at the ends so that they are held in place by the staple.

This section as a whole can form the basis of a technology unit by relating the science learned to everyday objects.

Sound Machines

 # How Does a Speaker Work?

Name: _____

Make sure the students do not touch any exposed wires.

- an exposed working speaker (for example, cut open the front speaker box of a radio set)
- salt

1. Lay down the radio so that the speaker is horizontal.
2. Sprinkle some salt on the speaker.
3. Turn on the radio on low volume.
   What happens to the salt?

   _____

4. Turn up the volume.
   What happens to the salt?

   _____

Speakers produce sound by vibrating.

Sound Machines

Make a Tin Can Telephone

Name: _____

- two tin cans with one end taken off each, and a hole through the bottom
- 5 yds. (5 m) of thin string (such as kite string)

1. Thread one end of the string through the hole in a tin can so the end is inside. Knot the end to secure it.
2. Thread the other end of the string through the hole in the other tin can so the end is inside. Knot the end to secure it.
3. Two students take a tin can each. They stand apart so that the string is pulled tight. One student speaks into his tin can and the other student listens in hers.

   Can you hear through the tin can? _____

   The sound waves travel from one can to the other through the string!

   Real phones work in the same way, except electricity carries the vibrations from one end to the other.

Sound Machines

Name: _____

Make a Pan Flute

Worksheet

- three straws
- scissors
- tape

1. Cut the straws in half.

2. Lay the straws next to each other. Tape them together.

3. Cut diagonally across the straws, so that you have short straws on the left and long straws on the right.

4. You can play your pan flute! Hold it close to your mouth, and blow into the straws.

Sound Machines

Name: _____

Make a Pan Flute

Worksheet

Which straws make high-pitched sounds?
_____

Which straws make low-pitched sounds?
_____

Pan was a mythical creature who played a pan flute. He had the bottom half of a goat and the top half of a man. Can you draw a picture of Pan playing his flute?

Sound Machines

Make a Guitar

- shoebox (no lid)
- four elastics
- stapler

1. Staple the elastics on one side of the shoebox.
2. Stretch one elastic as tight as you can across the open face of the shoebox. Staple it to the other side.
3. Repeat with the other elastics. Make each elastic a bit looser than the previous one.
4. You can play your guitar by flicking the elastics!

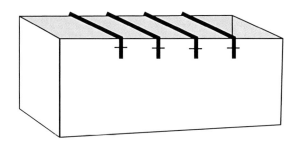

Name: _____

Make a Guitar

Worksheet

Which elastic makes the most high-pitched sound?

_____

Which elastic makes the most low-pitched sound?

_____

Listen carefully to the sound each elastic makes. Close your eyes. Ask a friend to flick one of the elastics. Write down which elastic you thought she flicked. Then, ask your friend which one she flicked. Were you right?

My guess	My friend's answer

Sound Crossword

ACROSS
2. The _____ inside our ear helps us hear sounds.
3. _____ cords in our throat produce sound.
5. Sounds can be loud or _____.
6. Vibrations produce _____.

DOWN
1. Sound _____ in all directions.
2. Bats use ____ to find insects.
4. A decibel meter measures how ____ sounds are.

Name: _____

Sound Wordsearch

```
S O U N L D V S I
E X P L O S I O N
A C A I U O B F T
R E D N D U R T E
D A P B E N A S N
R R I O F D T T S
U S T L O U I D I
M H C E C H O I T
G H H J L O N W Y
```

Eardrum
Ears
Echo
Explosion

Intensity
Loud
Pitch
Radio

Soft
Sound
Vibration

© On The Mark Press • S&S Learning Materials 92 OTM-2127 • SSB1-127 Light and Sound

Answer Key

Can You See in the Dark? (page 10)
What do you see in the dark? Nothing
Where is the ball? Answers will vary.
Can you see in the dark? No.
If you have no light source there is <u>darkness.</u>

The Source of Light (page 11)
Circled: TV, flashlight, lightning, candle
Not circled: Chicken, book, battery, lemon

Light Travels Fast! (page 13)
What do you see? The light appears on the ceiling at the same time you turn on the switch. Light travels too fast to see it move. How fast does light travel? Answers may vary. Light travels faster than any object.

Light Travels Straight (page 15)
In the drawing, beam will be straight.
Can you make light move? No.
Light travels straight.

You Light Up My Life (page 17)
Answers may vary.

Light-Detecting Plants (page 19)
Drawing should show that leaves change their position over time and turn to face sunlight. Plants do this because they need the light from the sun to grow.

How We Use Light Part 1 (page 20)

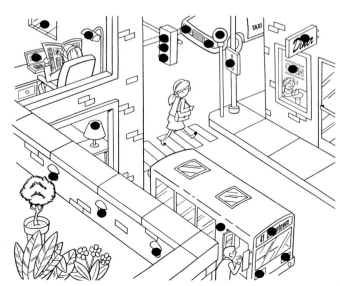

How We Use Light Part 2 (page 21)

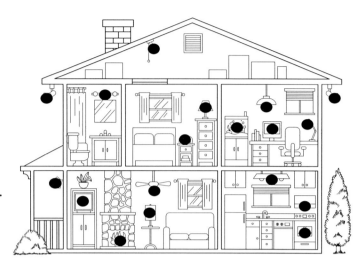

What Is a Light Bulb? Part1 (page 22)
1. No 2. No 3. Yes

What Is a Light Bulb? Part 2 (page 23)
1. No
2. Energy was supplied to the bulb and was turned into light
3. Answer could include any of: power, electricity, electrical energy, energy
4.

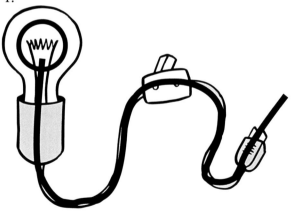

Do Light Bulbs Waste Energy? (page 25)
Answers will vary. The box will become hot. Students should conclude that light bulbs produce heat.

Does Light Spread Out? (page 29)
S is closest to the brightest area.
Light spreads out and becomes dimmer as it travels.

Brightness Decreases with Distance (page 31)
The number gauge will look dimmer and numbers will be harder to read as students move away. Checkmarks beside: The numbers became harder to read as I moved away from the light source; light gets dimmer as it moves away from its source.

How Does Light Spread Out? (page 33)
Answers will vary. As students move back, less sand will land in the cup. The sand is spreading out. As light moves away from its source, it spreads out.

Changing Shadows (page 36)
Answers will vary. The shape of shadows changes when the object is turned. Only a ball does not change because its shape always looks the same no matter how you rotate it. Objects make shadows when light shines on them from one direction.

Shadows and Light (page 38)
1. Shadows change shape and size when the direction of light changes. 2. At midday, shadows will be short, small. 3. In the morning, shadows will be long, big.

I See Through You! (page 40)
Answers may vary.

It's Play Time! (page 42)
1. translucent 2. opaque 3. shadow
4. Yes, they are larger and shape may change as characters are moved. 5. translucent

Bouncing Light (page 44)
Yes, light bounces.

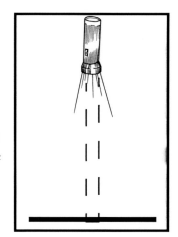 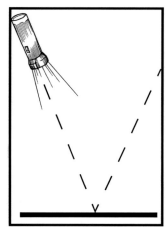

What Reflects Light? (page 46)
White paper reflects the most light, so more numbers are visible.

Absorbing Light (page 48)
The temperature will rise more in the black box as more light energy is absorbed. The temperature would not rise as much on a cloudy day, as the clouds are blocking the sun's light.

Where Does Sound Come From? (pages 50-52)
When you hit the drum, it vibrates, and the salt bounces around. A whistling sound happens when you blow on the whistle. There is a popping sound when you step on the balloon. Pictures should represent the experiment (drum, whistling, balloon popping) and ideally show the mechanism by which sound is produced in each case (vibration, blowing air, explosion).

Make A Kazoo (pages 54)
The kazoo makes sounds because the material vibrates. Answers will vary. Materials that vibrate work best.

Does Sound Travel? (pages 55-56)
1. A humming noise. 2. Ripples in the water.
3. Yes, the second glass makes the same sound as the first glass. 4. There are ripples in the water of the second glass. 5. TRUE, vibration and sound go together. 6. Draw a picture of the two

glasses with water: picture should represent the experiment. The arrow should go from the first to the second glass.

How Does Sound Travel? (page 58)
The Slinky™ goes from side to side too.

How Do Humans Make Sound? (page 60)

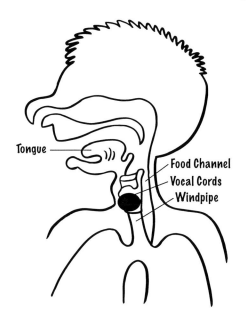

Make an Ear (page 62)
The wax paper vibrated and the salt bounced around on it.

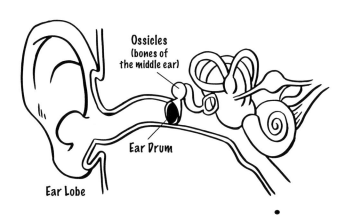

How Do Humans Hear Sounds? (page 63)

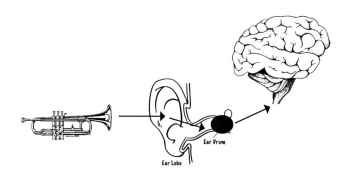

Why Do We Have Two Ears? (page 65)
Draw a picture of what you did: picture should reflect the group's experiment. Was the experiment easier with one or two ears? With two ears. Compare with your classmates. What did they say was easier? With two ears. Why do you think we have two ears? To make it easier to tell where sounds are coming from.

Speed of Sound (page 68)
1. One student walked to the other end of the field. 2. She blew the whistle and dropped her hand. 3. We saw the hand drop. 4. We heard the whistle.

Sound Travels (page 70)
1. Sound travels in all directions.
2. Everyone in the classroom.
3. Picture should show the experiment, with arrows pointing from the speaker towards all the students (in all directions).

Blocking Sound (page 72)
Which material will block the sound best? Answers will vary. Put a check mark in the column that tells what happened:

Material	Did not block	Blocked a bit	Blocked very well
Wood			X
Foam			X
Cardboard		X	
Tissue Paper	X		
Cloth		X	

Note: Answers may vary a bit depending on the exact materials used. The material that blocked the sound best was: answers may vary a bit depending on the exact materials used, but wood or foam is expected. The answer should be in keeping with the table shown.

Echoes and Bats (page 73)
Answers may vary.

Would You Be a Good Bat? (page 75)
Answers may vary.

Make a Xylophone (page 78)
Draw a picture of your ordered baby food jars: picture should show the baby food jars with ascending or descending water levels. The lowest-pitched sound came from the jar with: the most water. The highest-pitched sound came from the jar with: the least water.

How Loud Is Sound? (page 79)
What does a decibel meter measure? The decibel meter measures how loud a sound is.

The Intensity of Sound (page 80)
When you move away from the radio, the number it shows is smaller. The sound is less loud when you move away.

At the Zoo (page 81)
Answers may vary. For example:
High-pitched sounds: mice, bats, monkeys
Low-pitched sounds: frog, owl, lion
Soft sounds: snake, butterfly, bats
Loud sounds: elephant, monkey, lion

Sound Patrol (page 83)
Answers may vary depending on the objects tested. The conclusion should match the table.

How Does a Speaker Work? (page 85)
Turn on the radio on low volume. What happens to the salt? It bounces around a bit, or it doesn't bounce around at all. Turn up the volume. What happens to the salt? It bounces around a lot.

Make a Pan Flute (page 88)
The shortest straws make high-pitched sounds. The longest straws make low-pitched sounds.

Make a Guitar (page 90)
The tightest elastic makes the highest-pitched sound; the loosest elastic makes the lowest pitched sound.

Sound Crossword (page 91)

Sound Wordsearch (page 92)

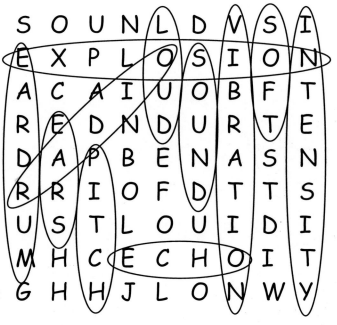